Tractors

Angela Royston

Illustrated by
Terry Gabbey

Heinemann Interactive Library
Des Plaines, Illinois

Contents

Tractors 4

Early Tractors 6

Hay-making 8

Driving a Tractor 10

Plowing 12

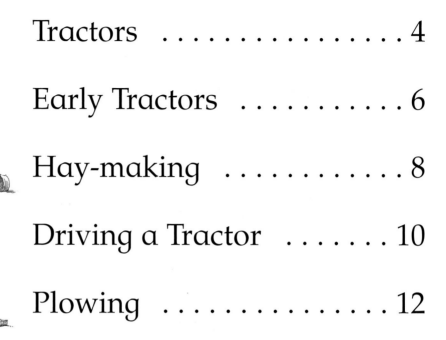

© 1998 Reed Educational & Professional Publishing
Published by Heinemann Interactive Library, an imprint of Reed Educational & Professional Publishing,
1350 East Touhy Avenue, Suite 240 West
Des Plaines, Illinois 60018

Printed and bound in Italy. See-through pages printed by SMIC, France.

Library of Congress Cataloging-in-Publication Data
Royston, Angela.
 Tractors / Angela Royston ; illustrated by Terry Gabbey.
  p.   cm. — (Inside and out)
 Includes bibliographical references (p.   ) and index.
 Summary: Introduces different types of tractors and their uses.
 ISBN 1-57572-180-5 (lib. bdg.)
 1. Tractors—Juvenile literature.   [1. Tractors.]
 I. Gabbey, Terry, ill. II. Title. III. Series.
 TL233.R67   1998
 629.225'23—dc21   97-41189
                                                        CIP
                                                         AC

Acknowledgements
The Publishers would like to thank the following for permission to reproduce photographs: page 5: Art Directors Photo Library © Trip/D Houghton;
page 7: Tony Stone Worldwide © Bruce Hands; page 8: Tony Stone Images © Peter Dean; page 15: © Britstock-IFA/Frisch; page 17: © Christine Osborne Pictures;
page 19 top right: © ZEFA-CPA; page 19 bottom right: ZEFA © Smith R.; page 21: Tony Stone Images © Mitch Kezar; page 23: © Bomford Turner.

Some words are shown in bold, **like this**.
You can find out what they mean by looking in the glossary.

02 01 00 99 98
10 9 8 7 6 5 4 3 2 1

Planting Seeds 14

 Spraying Crops 16

 Harvesting Crops 18

 Combine Harvesters 20

 All Kinds of Jobs 22

 *Index and Glossary* 24

More Books to Read 24

Planting Seeds 14

Spraying Crops 16

Harvesting Crops 18

Combine Harvesters 20

All Kinds of Jobs 22

Index and Glossary *24*

More Books to Read *24*

Tractors

Tractors do more jobs on the farm than any other machine. They pull trailers and **plows,** and have attachments to cut bushes and spray crops. They can be driven over rocky ground and through deep mud.

Most tractors have big wheels with huge tires. This tractor has tracks instead of wheels. It drives over sharp stones that would rip up rubber tires.

Not all tractors are used on farms. This tractor is towing a boat out of the sea. The boat is on a trailer.

Early Tractors

Some of the first farm machines were powered by steam. This steam engine was used more than 100 years ago to shake seeds of grain off stalks of wheat. It was later replaced by the combine harvester.

This early tractor was designed in 1908 by Henry Ford. He called tractors "automobile **plows**." Ford also **invented** the first low-priced car.

Before tractors were invented, horses and other animals pulled farm machines. This **Amish** boy in Pennsylvania still uses five horses to pull his plow.

Hay-making

Grass grows quickly in spring. By May it is ready to be cut. The tractor pulls the mowing machine over the field. It cuts the grass and then leaves it to dry to become hay.

Some farmers do not wait for the grass to dry. This machine chops the freshly cut grass and blows it into a trailer. The wet grass is stored as silage.

Silage is stored and fed to cows and other animals in the winter when there is little grass for them to eat.

The farmer then brings a baling machine to the hay-field. It gathers and presses the hay into big bales. What is the other tractor doing?

Driving a Tractor

Inside the **cab** of a tractor there are many levers and switches. The farmer moves these to work whatever machine or attachment the tractor is using. A heater keeps the cab warm in winter, and an air conditioner keeps it cool in summer.

This tractor is pulling a car out of the snow. A light flashes on top of the cab to warn other drivers that the tractor is there.

The driver of this tractor has attached a special blade to the front of it. The blade pushes the snow to the side of the road. Now the road is clear for cars.

Plowing

Before they can plant seeds, farmers prepare the land. The soil must be turned over to make it easier for the seeds to grow. This tractor pulls a **plow** over the field. The blades dig deep into the soil and turn it over.

Watch out! This farmer is spreading **manure** and straw over the land. The manure fertilizes the soil and makes the seeds grow better.

The round blades of this harrow behind this tractor break up the soil into smaller pieces. Now the farmer can plant the seeds.

Planting Seeds

Seeds are carried in the flat box behind the tractor. The sharp spikes of the **drill** cut long, narrow holes in the soil. As the seeds drop into the holes, the machine covers them with soil.

These three women are planting lettuce. They push the seedlings into the soil as the tractor moves slowly across the field.

Seeds and plants need plenty of water to grow. A tractor has pulled a pump and hosepipe into a field to water the crop.

Spraying Crops

Sometimes insects or diseases attack a farmer's crop. Spraying special chemicals on the crops will protect them. A sprayer is hooked onto each tractor and liquid chemicals are poured into the sprayer's tank.